4th Grade Science Volume 3

© 2013 Todd Deluca
OnBoard Academics, Inc
Newburyport, MA 01950

800-596-3175
www.onboardacademics.com

Table of Contents

Skeletal and Muscular Systems

Your Skeletal System's Many Functions

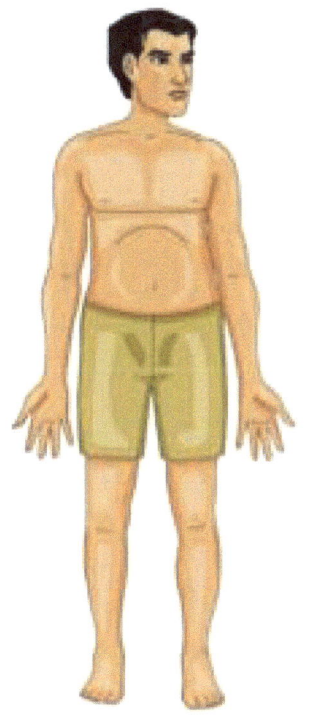 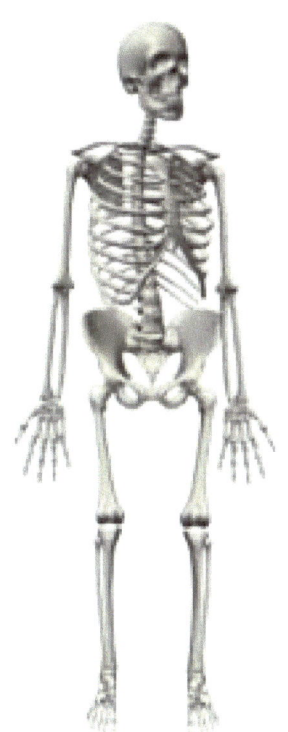

☐ stores minerals

☐ makes blood

☐ provides support

☐ protects your body

☐ allows movement

The Skeletal System

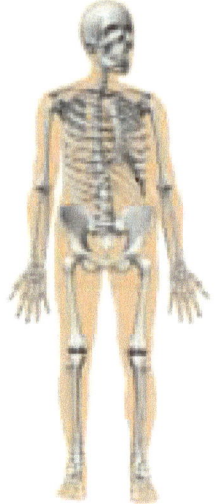

The skeletal system is the name we give to your bones and the tissues such as tendons and ligaments that connect your bones to each other and to your muscles.

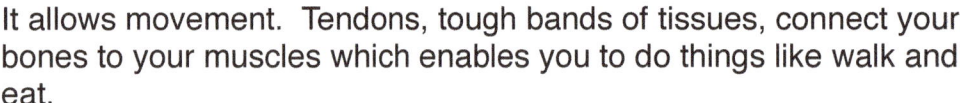

The skeletal system has five main functions: it provides support for your body, it protects your body, it allows movement, it makes blood, and it stores important minerals.

The skeletal system supports your body. If you didn't have a skeleton, your body would be a floppy blob.

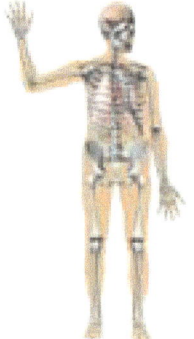

It protects your body. The 206 bones that make up your skeleton act like a suit of armor to protect the organs in your body such as your brain, your heart and your lungs.

It allows movement. Tendons, tough bands of tissues, connect your bones to your muscles which enables you to do things like walk and eat.

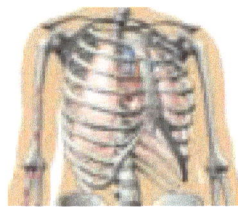

It makes blood. You might me be surprised to learn that your heart does not make blood. Your heart is a muscle that serves to pump the blood throughout your body. Blood cells are made in marrow which is a jelly like substance found inside your bones.

Bones in your skeleton also act as a storage area for important minerals that you body needs such as phosphorus and calcium.

If you thought your bones were a nonliving part of your body you will be surprised to learn that bones are very much alive with their own nerves and blood vessels.

Name that bone.

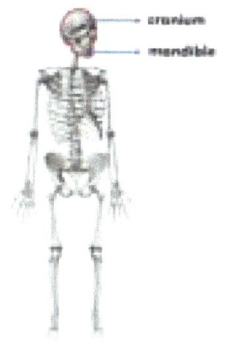
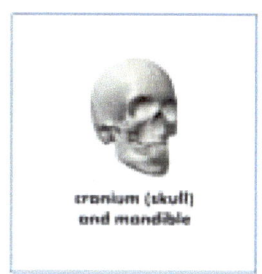

cranium (skull)
and mandible

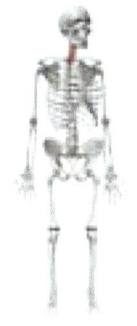
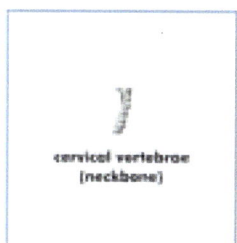

cervical vertebrae
(neckbone)

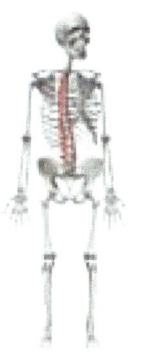
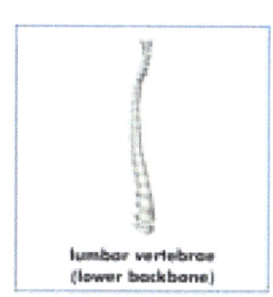

lumbar vertebrae
(lower backbone)

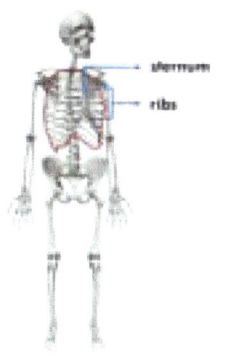
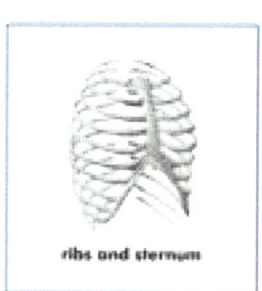

ribs and sternum

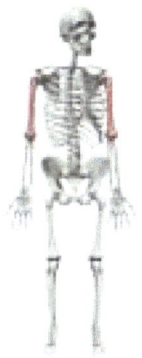
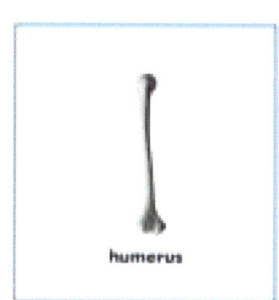

humerus

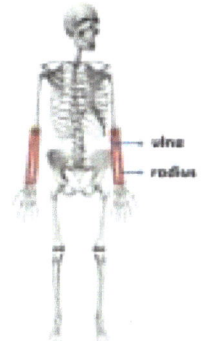
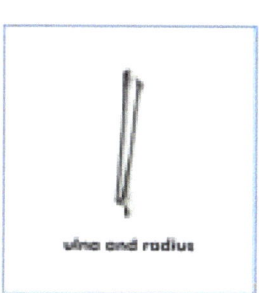

ulna and radius

More name that bone.

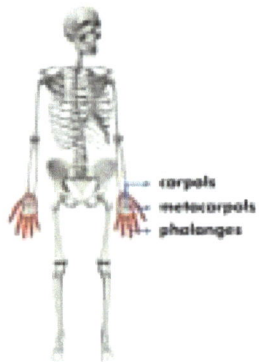 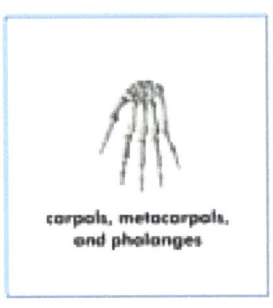

carpals, metacarpals, and phalanges

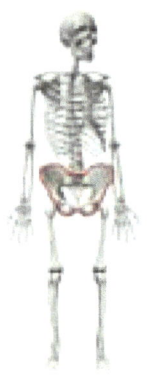 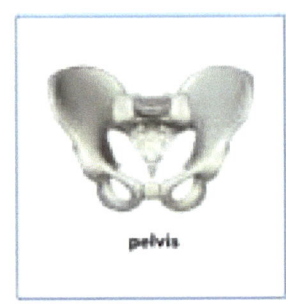

pelvis

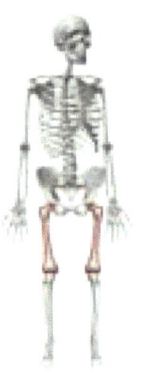 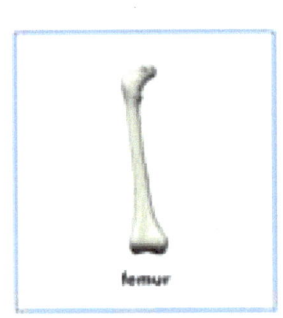

femur

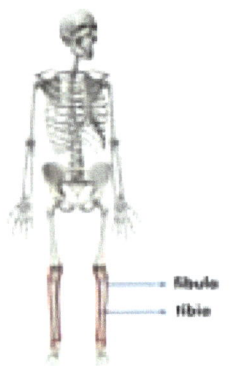 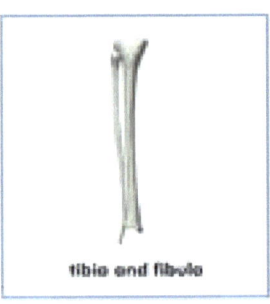

tibia and fibula

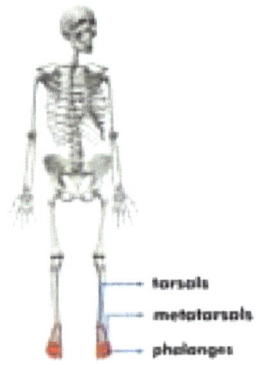 

tarsals, metatarsals, and phalanges

OK Dr. Frankenstein, put this skeleton back together and label the bones.

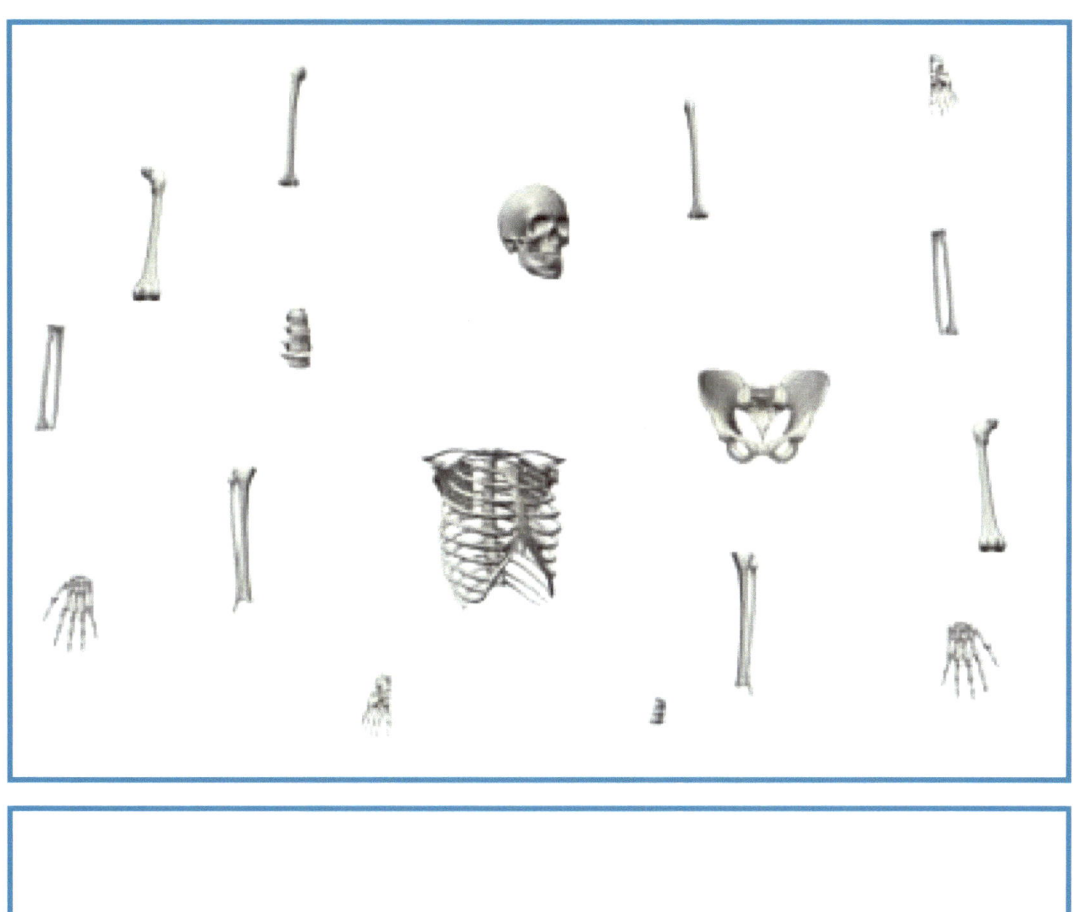

Which bone am I?

Read the definition and then match it to the proper bone.

phalanges	I'm the part of skull that encloses the brain.
cranium	I go from the shoulder to the elbow.
fibula	I am the bones that form the fingers and toes.
sternum	I am sometimes called the breast bone.
humerus	I am sometimes called the thigh bone and am the largest and strongest gone in the body.
femur	I am the thin bone on the outside of your leg.

Skeletal System Quiz

1. The human skeleton is composed of _____ bones.
    a. 206
    b. 106
    c. 306

2. The skeleton gives your body its shape. What will happen your body does not have a skeleton? _____
    a. The body would be unable to move.
    b. You would be a big blob.
    c. Your body would lose protection.
    d. All of the above.

3. The bones are alive with nerves and blood vessels. True or false?

4. The skeletal system does all of these functions except: _____
    a. Provide support
    b. Allow movement
    c. Protect internal organs
    d. Help digestion

5. I go from hip to knee. What am I? _____

The Cardiovascular System

The Heart; 360°

Front

Right

Rear

Left

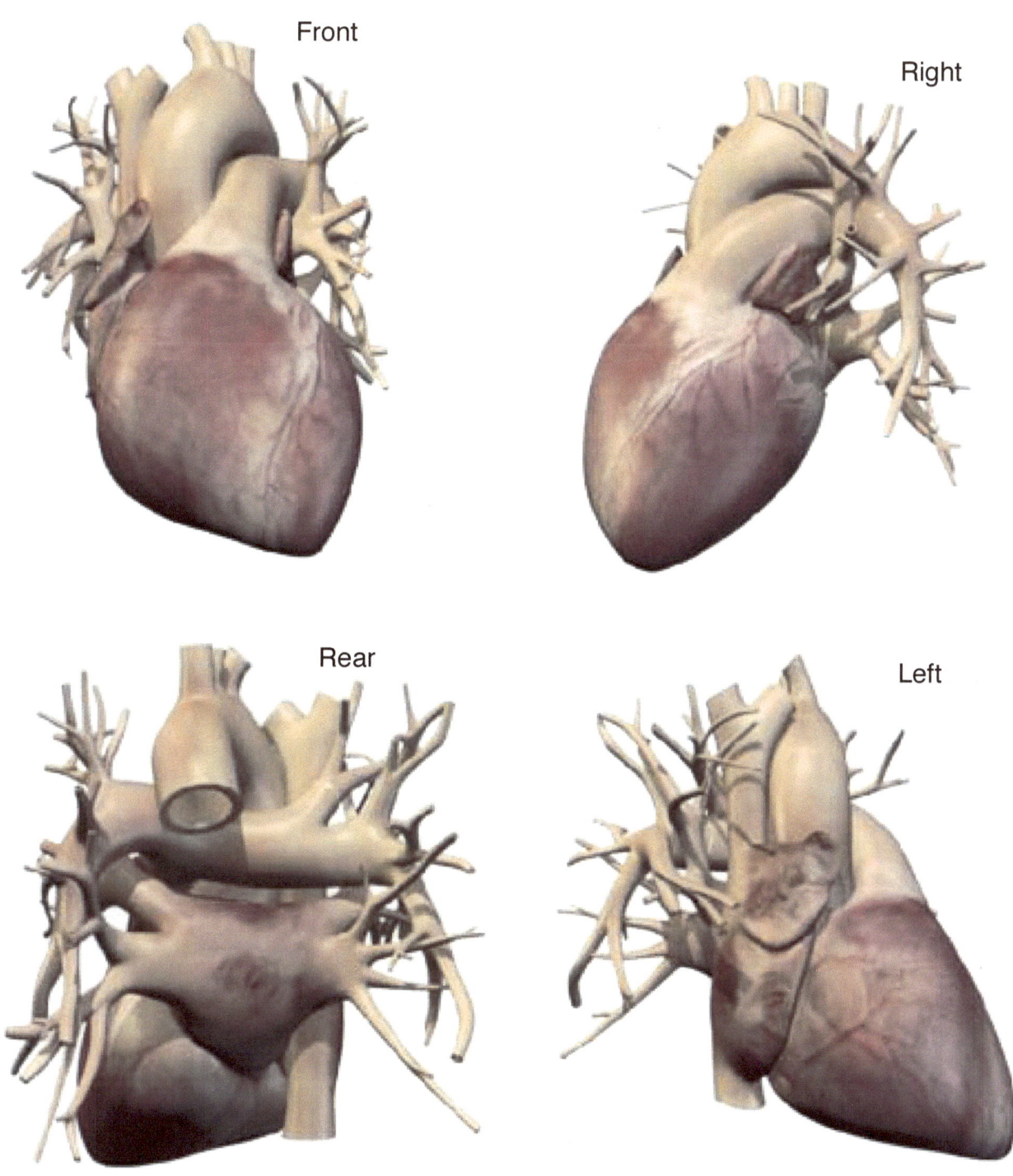

Location of the heart.

Draw a circle or a heart in the area of the body where the heart is located.

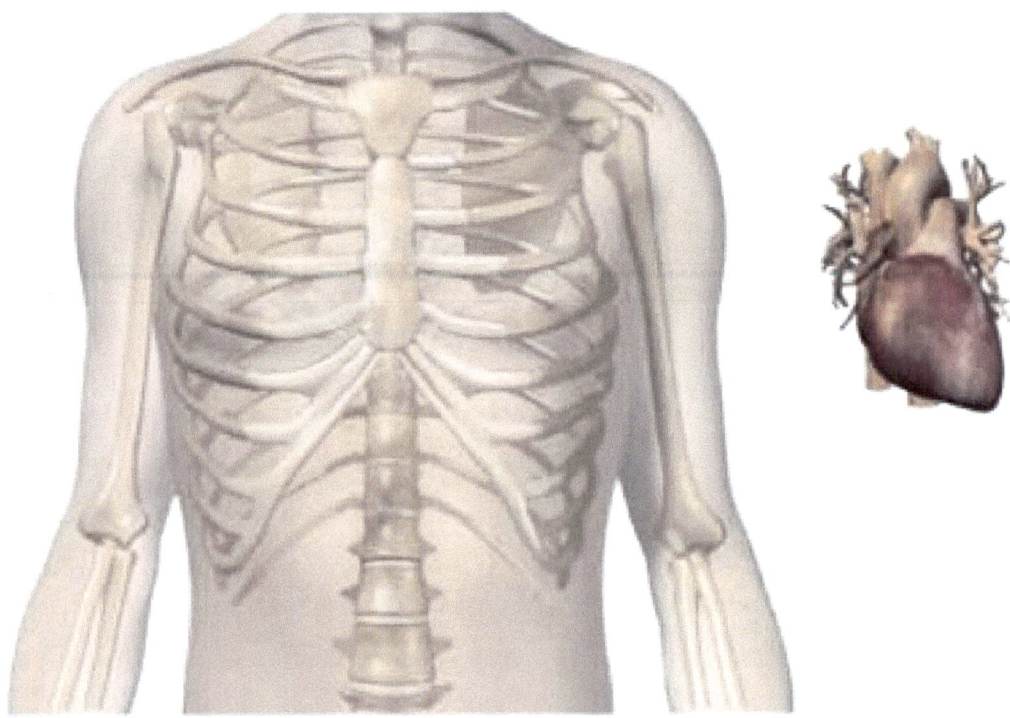

How fast does the heart beat?

heart rate

70 beats per minute

Stand

heart rate

80 beats per minute

Walk

heart rate

100 beats per minute

Jog

heart rate

120 beats per minute

Run

> **The rate at which the heart beats varies between people of different ages and different levels of fitness, but the heart always beats faster when we do exercise. Why do you think this is?**

_____

_____

_____

_____

The Heart and the Cardiovascular System

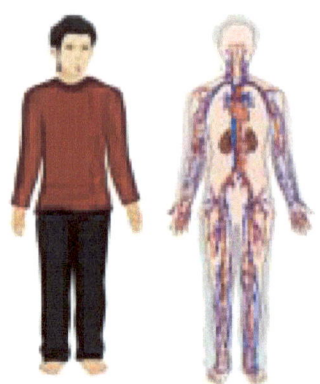

The cardiovascular system is the name that we give to the network of organs and pipes that deliver oxygen and nutrients to all of the cells within our body. The pipes in this system are called blood vessels and the oxygen and nutrients are transported to our cells in the form of blood.

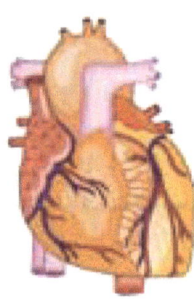

At the center of our cardiovascular system is our heart, a pear shaped muscle that pumps our blood 24 hours a day, every single day of our lives without a break. Let's take a closer look at the heart to understand how it works.

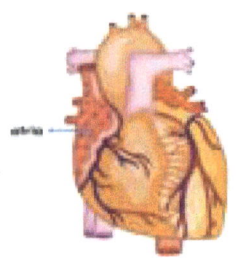 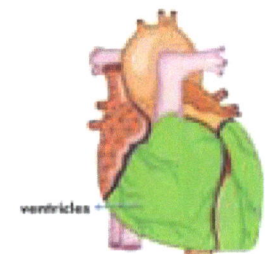

The heart has four compartments called chambers. The two chambers at the top are called atria and the two chambers at the bottom are called ventricles The heart works by contracting and relaxing. When the heart contracts, blood is squeezed from the ventricles in to large arteries.

When the heart relaxes, this allows blood to flow into the atria from veins. The atria store the blood until its ready to be pumped out by the ventricle. Valves within the four chambers of the heart open and close to facilitate flow in and out of the heart.

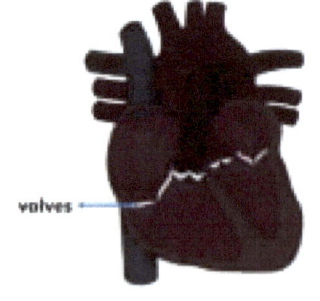

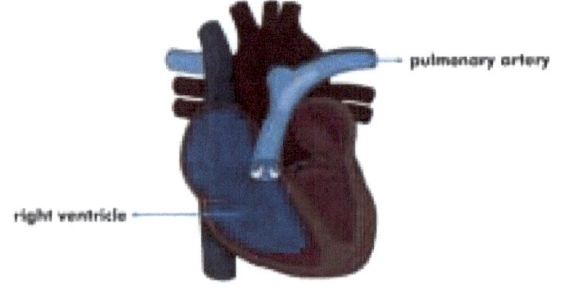

The right side of the heart (left side on the illustration) pumps the oxygen into the lungs. Deoxygenated blood means that the blood does not have any oxygen in it. In our illustrations, this is represented as blue. The deoxygenated blood travels from the right ventricle through the pulmonary artery and into the lungs where it absorbs oxygen and gets rid of unwanted carbon dioxide.

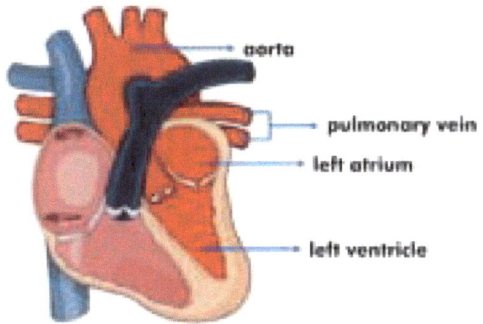

The oxygenated blood, represented by the color red, then flows back to the heart through the pulmonary veins into the left atrium. Its then pumped by the left ventricle through the aorta, the largest artery in the body, and by way of a network of increasingly small blood vessels on to all of the cells in your body where it distributes fresh oxygen.

We call the flow of blood to and from the lungs, the pulmonary system and the blood flow to and from the rest of the body the systemic system.

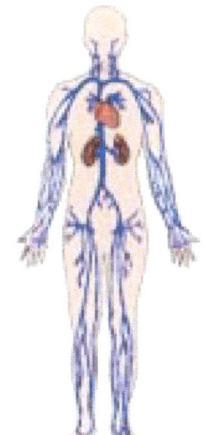

Within the systemic system, when the blood reaches the cells the oxygen and nutrients are exchanged for waste products such as CO_2 which the body can't use. The deoxygenated blood takes a return trip back to the heart where it enters the right atrium through two large veins called the superior and inferior vena cava and then the whole cycle repeats.

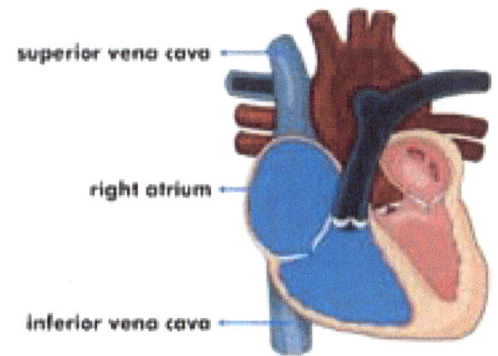

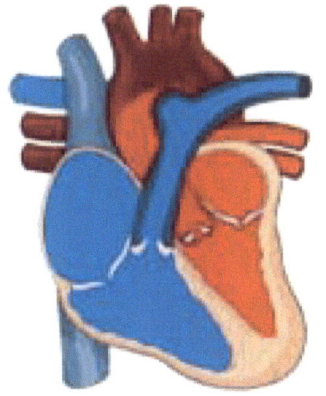

Amazingly one complete cycle only takes about 30 seconds. A heart beat consists of one contraction and one relaxation. In the previous exercise we learned that when we exercise our heat beats or heart rate increases. That's because our cells need more oxygen and generate more waste. This means that our heart must beat faster to increase the distribution of blood to and from the cells.

Order these steps in the cardiovascular system by labeling them 1-8.
The first step has been entered for you to help to get started.

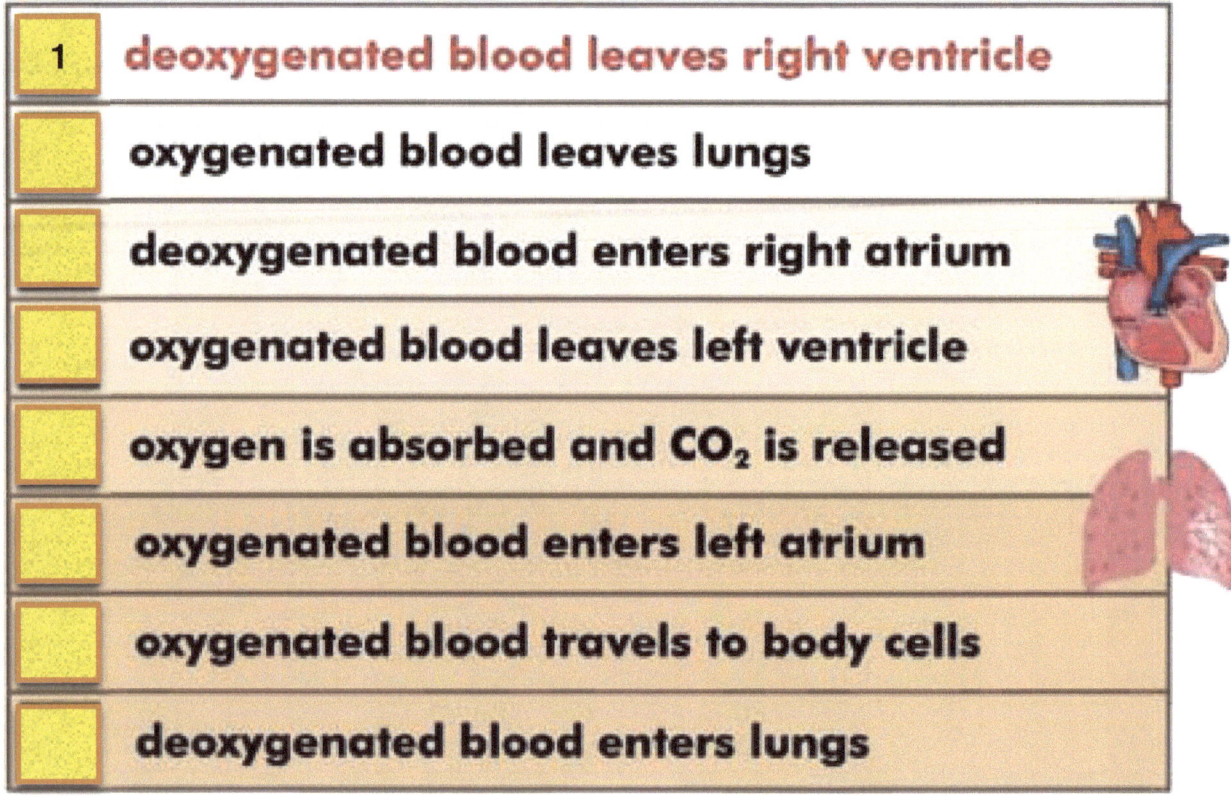

1	deoxygenated blood leaves right ventricle
	oxygenated blood leaves lungs
	deoxygenated blood enters right atrium
	oxygenated blood leaves left ventricle
	oxygen is absorbed and CO_2 is released
	oxygenated blood enters left atrium
	oxygenated blood travels to body cells
	deoxygenated blood enters lungs

Label the great vessels of the heart.

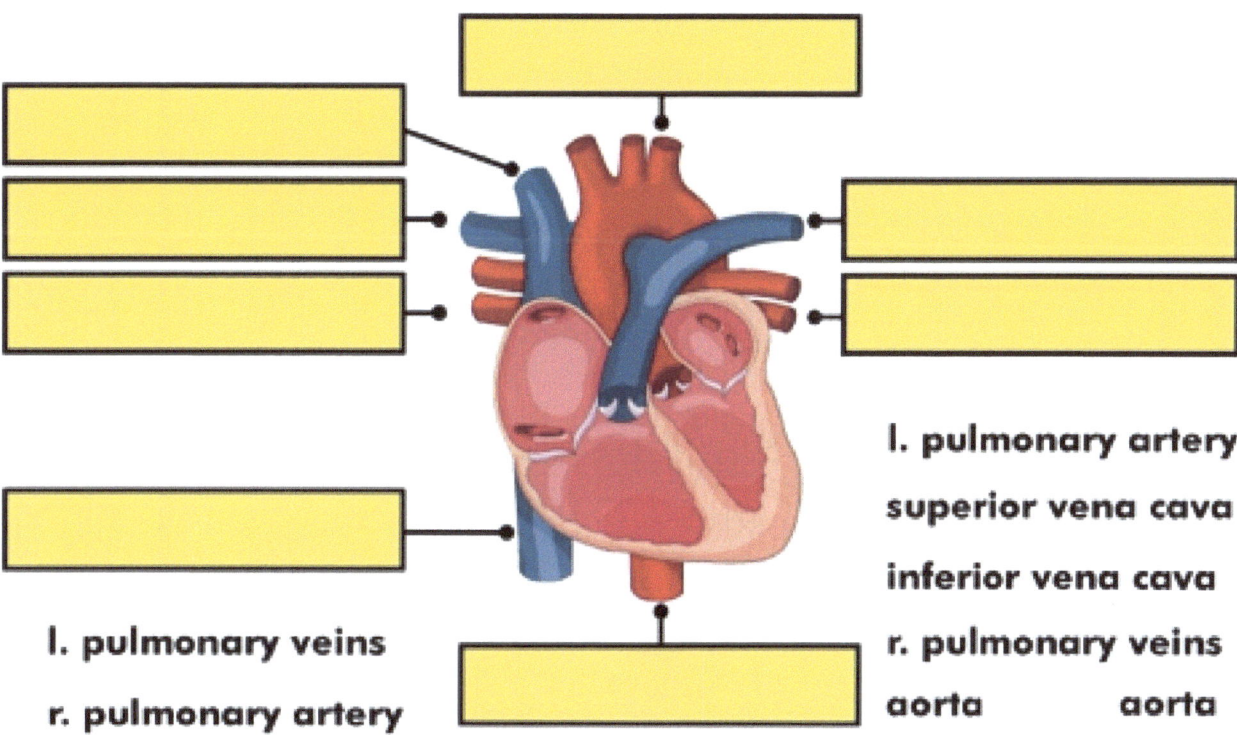

l. pulmonary veins

r. pulmonary artery

l. pulmonary artery

superior vena cava

inferior vena cava

r. pulmonary veins

aorta   aorta

The Cardiovascular System Quiz

1. The pipes in the cardiovascular system are called
   _____.
   - a. wires
   - b. ventricles
   - c. atria
   - d. blood vessels

2. There are _____ chambers in the heart.

3. The right side of the heart pumps oxygenated blood to the lungs. True or false?

4. In the pulmonary system, blood flows _____.
   - a. to and from the lungs
   - b. to and from the heart

5. Superior and inferior vena cava are two large veins present in the _____.
   - a. left ventricle
   - b. left atrium
   - c. right ventricle
   - d. right atrium

The Digestive System

What is the length of your digestive system?

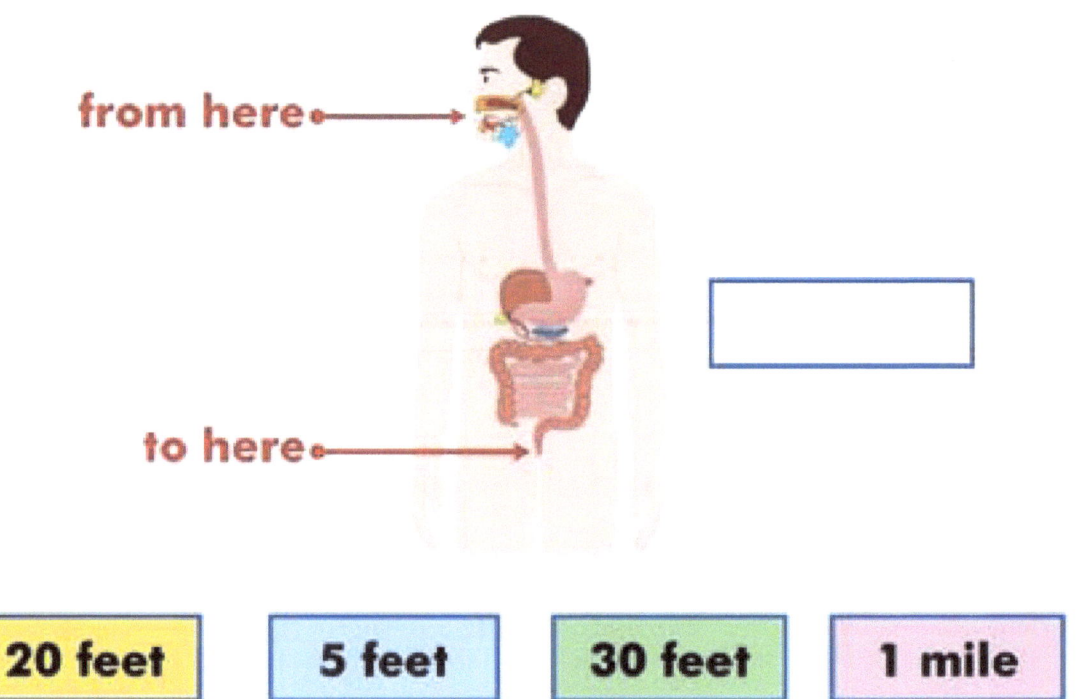

from here

to here

| 20 feet | 5 feet | 30 feet | 1 mile |

The digestive system's job is to break down everything we eat and drink, and to convert it into nutrients that the body can use. The digestive tract (the organs and tubes through which food and liquid passes) begins with the mouth and ends with the anus, and is 30 feet long, 20 feet of which is the small intestine.

What happens to the food we eat?

Digestion begins in the mouth where food is broken down by the teeth and chemicals in the saliva. The broken down food called the bolus travels down the esophagus and into the stomach. The stomach churns and breaks the bolus down into particles with the help of acids. It then pushes these particles into the duodenum which is the first part of the small intestine. In this part of the system the particles are broken down further with the help of chemicals supplied by the liver and pancreas. The particles pass through the rest of the small intestines where the food is absorbed into the blood and passed throughout the body to be used as energy. The food that is not absorbed passes on to the large intestine where water is extracted. The parts that can't be absorbed are eliminated through the rectum and from there as waste through the anus.

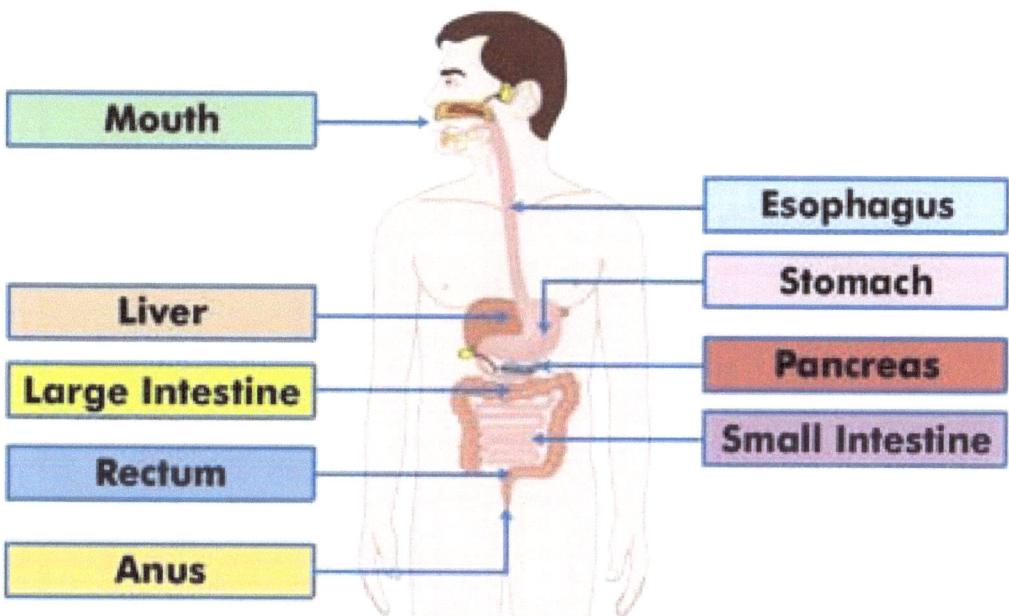

Draw the route that food takes through the body.

Order the Organs!
In which order does food pass through your digestive system?

| Pancreas |
| Large Intestine |
| Esophagus |
| Stomach |
| Mouth |
| Rectum |
| Anus |
| Small Intestine |
| Liver |

Two of the organs are accessory organs. They produce chemicals which help us to digest food, but food doesn't pass through these organs.

Label the parts of the digestive system.

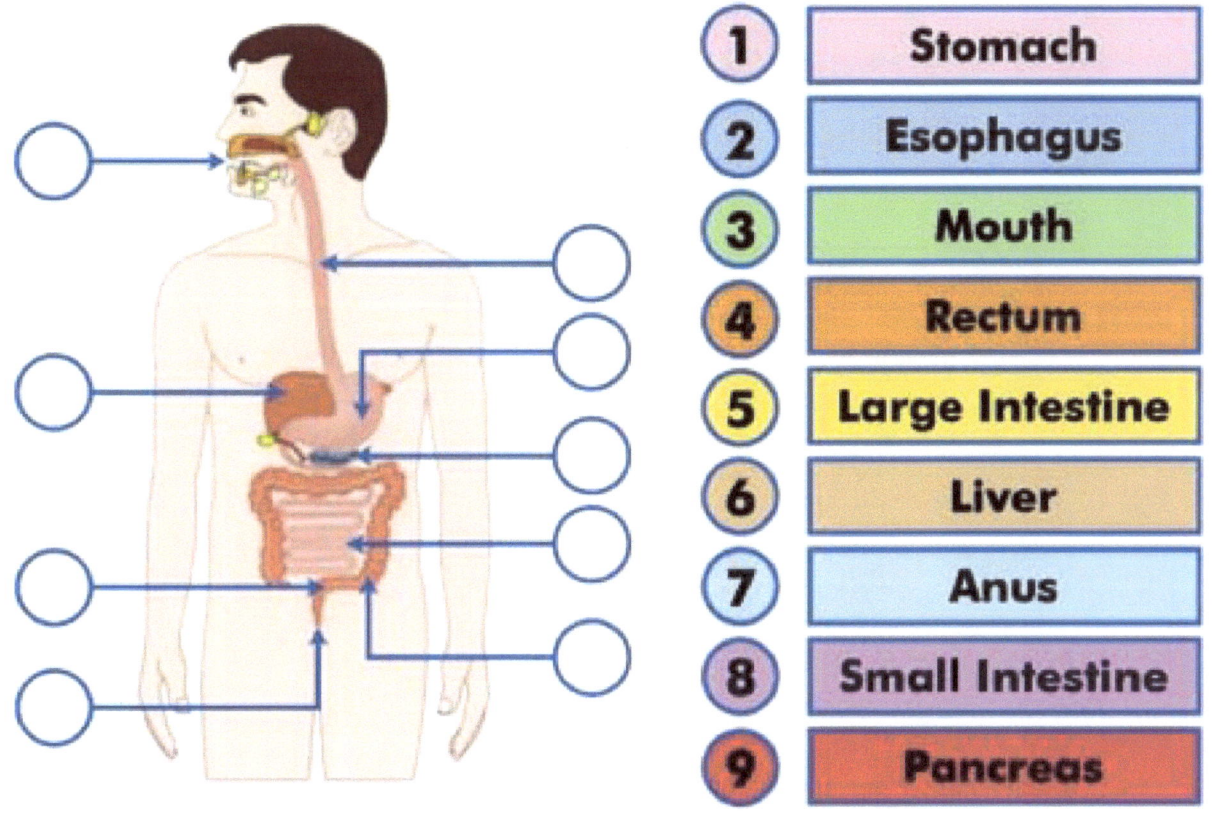

1. Stomach
2. Esophagus
3. Mouth
4. Rectum
5. Large Intestine
6. Liver
7. Anus
8. Small Intestine
9. Pancreas

Which organ am I?

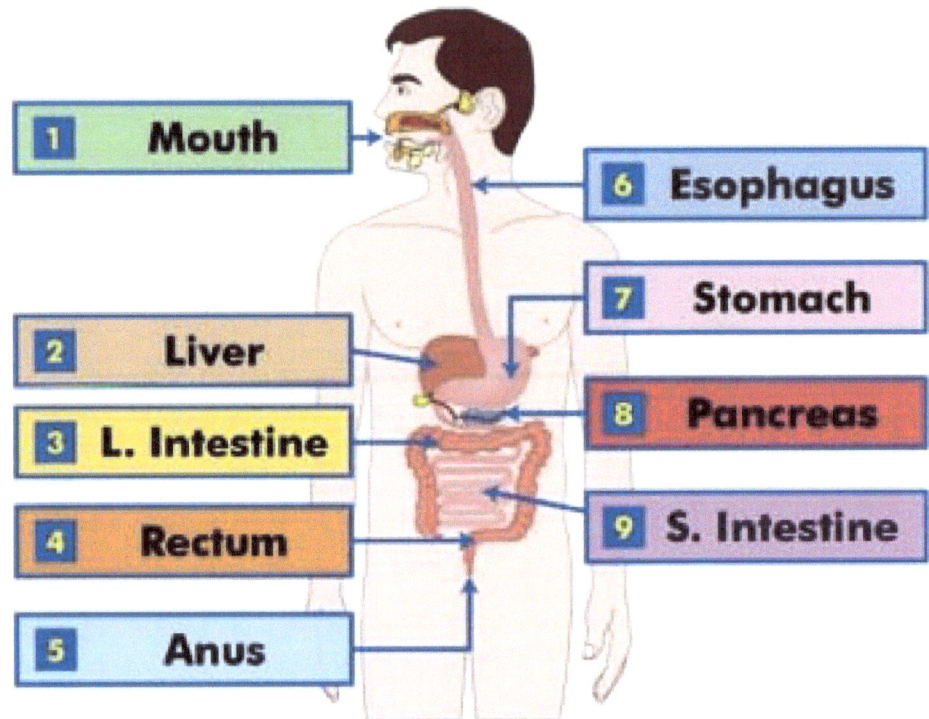

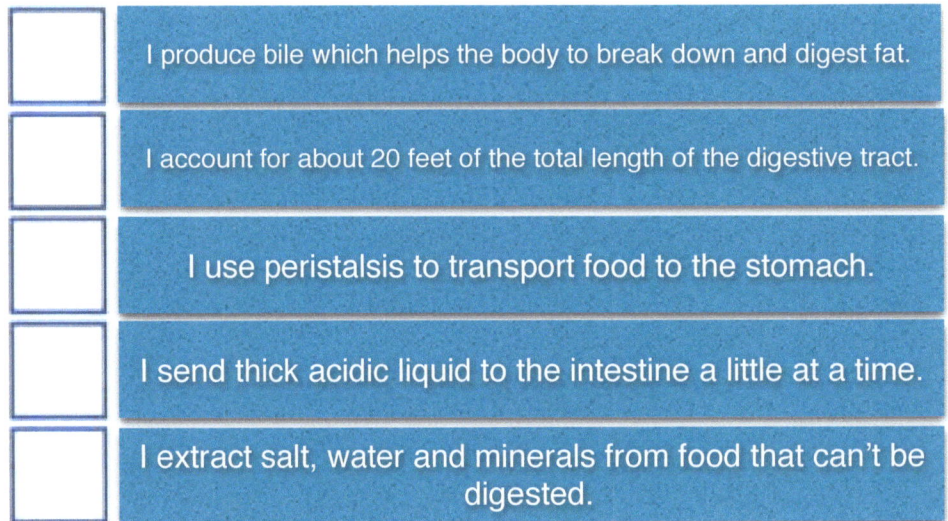

	I produce bile which helps the body to break down and digest fat.
	I account for about 20 feet of the total length of the digestive tract.
	I use peristalsis to transport food to the stomach.
	I send thick acidic liquid to the intestine a little at a time.
	I extract salt, water and minerals from food that can't be digested.

The Digestive System Quiz

1. Digestion begins in the _____.

2. Food is mashed in the mouth with the help of _____.
    a. chyme
    b. bolus
    c. saliva

3. The long pipe that connects the mouth with the stomach is called the _____.
    a. pharynx
    b. small intestine
    c. large intestine
    d. esophagus

4. The acids present in the stomach break down the food into smaller particles. True or false?

5. The nutrients from the food we eat are absorbed into the _____.
    a. stomach
    b. small intestine
    c. large intestine
    d. liver

The Respiratory System

How much air (in liters) do you think your lungs can hold?
Think about your answer and write half of the total on each lung.

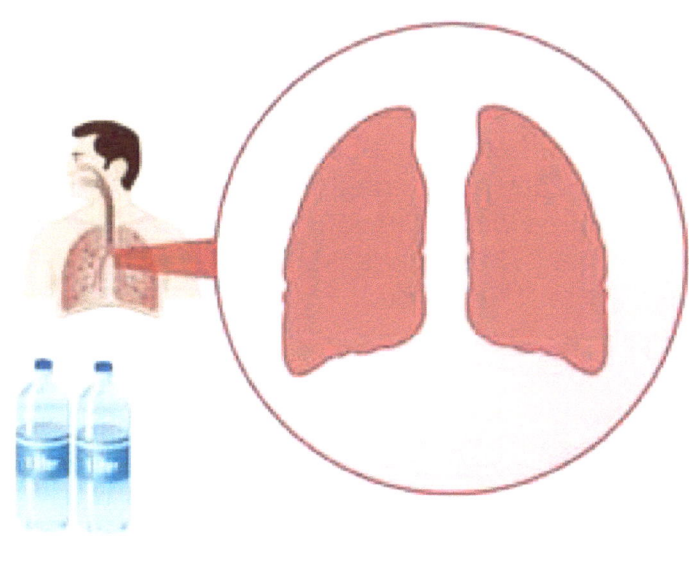

_____

What's in the air we breathe?

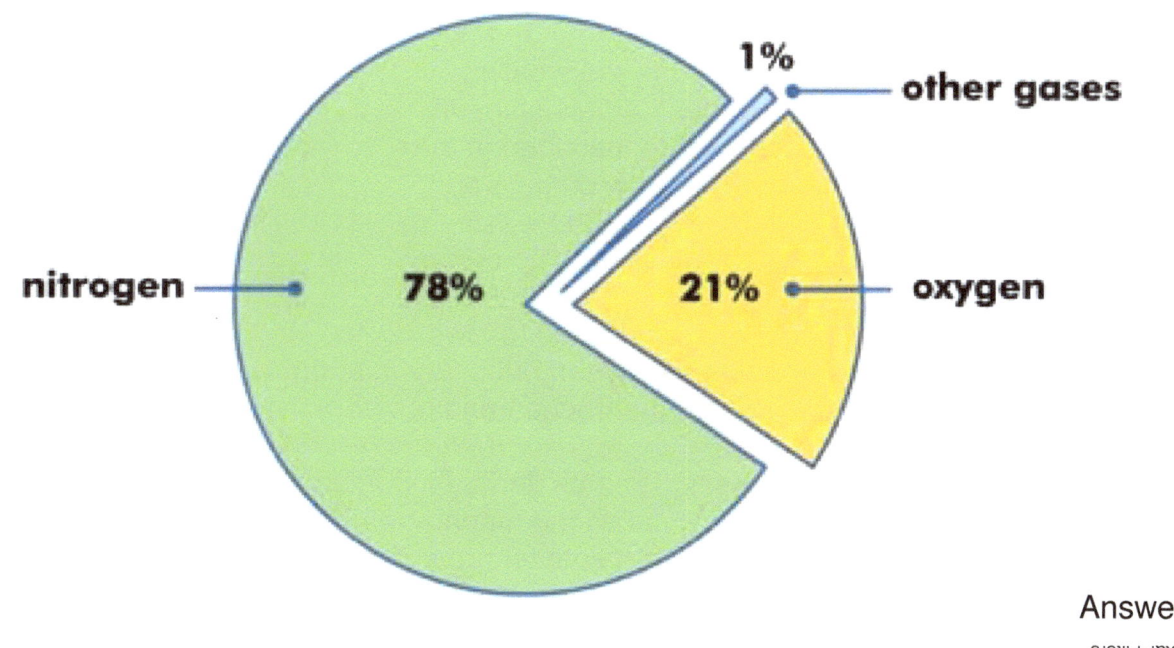

Answer

2 liters per lung
Total 4 liters

What happens to the air we breathe?

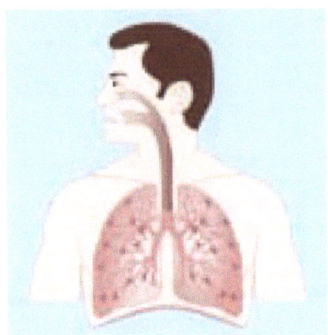

Every cell in your body needs oxygen to release energy from the food that you eat. Without oxygen, you body would simply not work.

Oxygen enters the body through the nose and mouth and then flows down through the trachea located in your throat. The trachea is also sometimes called the windpipe. The trachea separates into two paths called the bronchial tubes. The bronchial tubes mark the beginning of the lungs. The lungs are some of the largest organs in the body and have a spongy consistency.

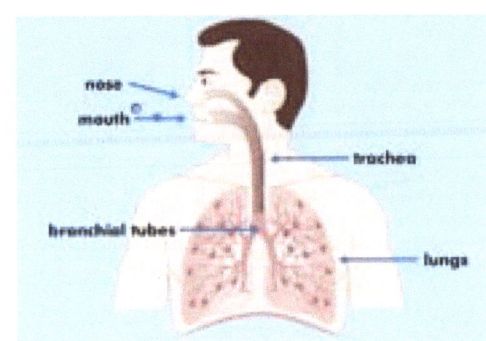

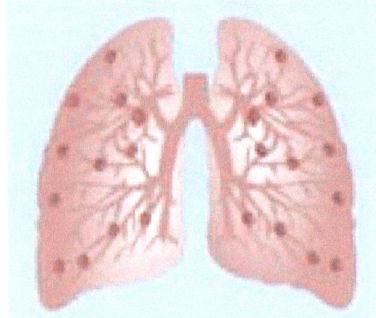

Within the lungs the brachial tubes separate into narrower and narrower tubes much like the branches in a tree.

The tiny branches end with tiny sacs called alveoli. Alveoli are surrounded by tiny blood vessels called capillaries. At this stage the oxygen in the alveoli gets absorbed into the blood in the capillaries. At the same time there is an exchange as the blood releases carbon dioxide which is the waste product your body creates when it turns food into energy.

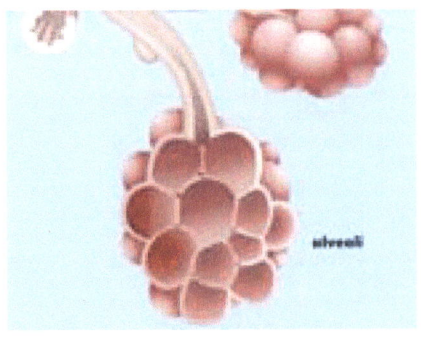

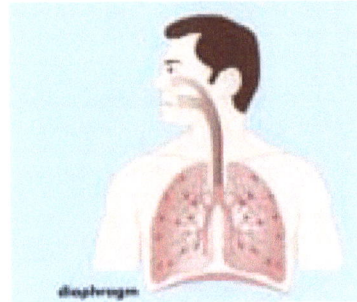

The carbon dioxide follows the flow of oxygen but in reverse; through the bronchial tubes, up the trachea and out from the nose and mouth.

Breathing is controlled by a domed shaped muscle beneath the lungs called the diaphragm. When the diaphragm contracts, it creates more space for the lungs so the can expand and suck in more air. When the diaphragm relaxes, there is less space for the lungs so air is forced out.

Label the parts of the respiratory system.

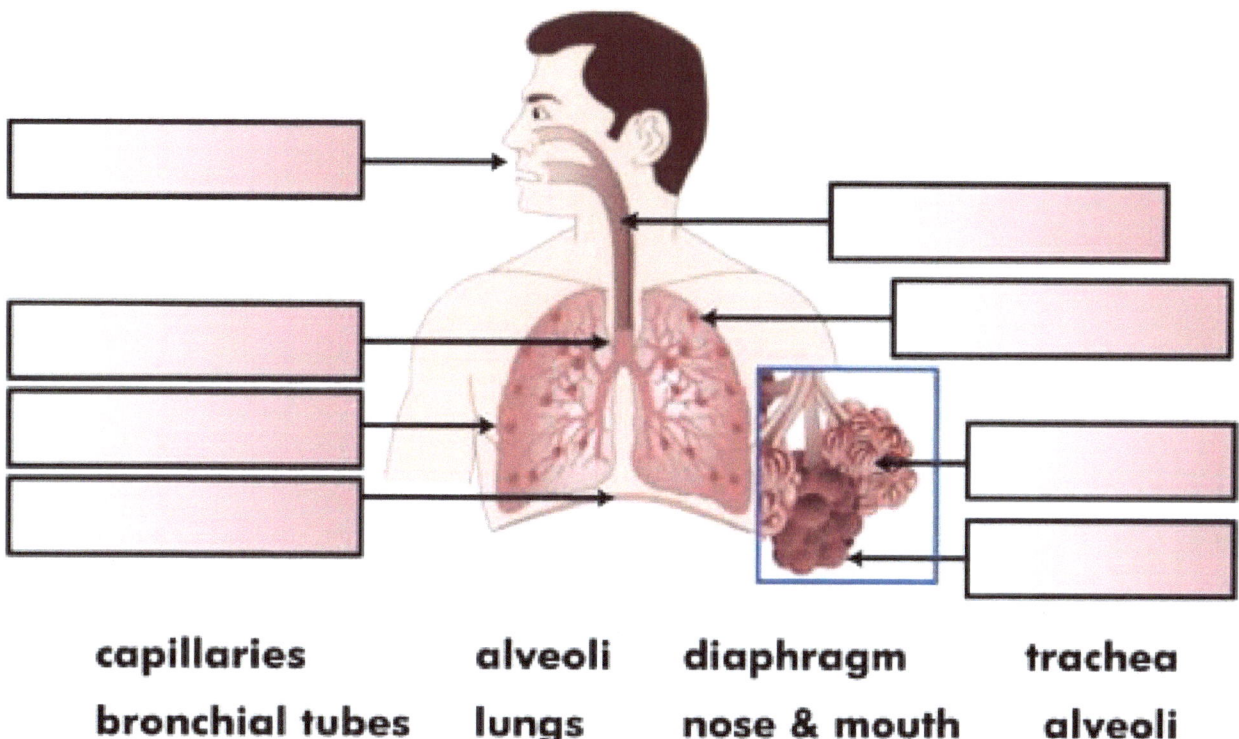

capillaries   **alveoli**   **diaphragm**   **trachea**

bronchial tubes   **lungs**   **nose & mouth**   **alveoli**

Put these breathing steps in order.

☐	oxygen passes through bronchial tubes
☐	cells use oxygen for energy; create CO_2 as waste
☐	oxygen absorbed by capillaries
☐	diaphragm contracts
☐	oxygen passes through trachea
☐	oxygen enters alveoli
☐	oxygen enters nose and mouth
☐	carbon dioxide is exhaled when diaphragm relaxes
☐	blood delivers oxygen to cells

1	2	3
4	5	6
7	8	9

Which part of the respiratory system am I?

Match the term with the correct descriptions.

I am the gateway that allows oxygen into the body.		**trachea**
		alveoli
I am a tiny bubble like sac where oxygen and carbon dioxide get exchanged.		**lungs**
I am the muscle that makes breathing possible.		**capillaries**
I am a spongy organ that divides into millions of tiny branches.		**diaphragm**
		bronchial tubes
I am also known as the windpipe.		**nose & mouth**

The Respiratory System Quiz

1. The human lungs can hold 10 liters of oxygen? True or false?

2. Our atmosphere is primarily _____ gas.
    a. Hydrogen
    b. Oxygen
    c. Nitrogen

3. Oxygen enters the body through a tube called the
    a. trachea
    b. bronchial tubes
    c. alveoli

4. The trachea separates into two paths called _____.
    a. windpipes
    b. bronchial tubes
    c. capillaries

5. Tiny sacs at the end of the branches of bronchial tubes are called alveoli. True or false?

6. Exchange of oxygen and carbon dioxide happens in the alveoli. True or false?

Newburyport, MA 01950

1-800-596-3175

OnBoard Academics employs teachers to make lessons for teachers! We create and publish a wide range of aligned lessons in math, science and ELA for use on most EdTech devices including whiteboard, tablets, computers and pdfs for printing.

All of our lessons are aligned to the common core, the Next Generation Science Standards and all state standards.

If you like our products please visit our website for information on individual lessons, teachers licenses, building licenses, district licenses and subscriptions.

Thank you for using OnBoard Academic products.